Rocks and Soil

written by Maria Gordon
and
illustrated by Mike Gordon

Wayland

Simple Science

Air
Colour
Day and Night
Electricity and Magnetism
Float and Sink
Heat
Light
Materials
Push and Pull
Rocks and Soil
Skeletons and Movement
Sound

Series Editor: Catherine Baxter
Advice given by Audrey Randall – member of the Science Working Group for the National Curriculum.

First published in 1996 by
Wayland (Publishers) Ltd
61 Western Road, Hove
East Sussex, BN3 1JD, England

British Library Cataloguing in Publication Data
Gordon, Maria
Rocks and Soils. - (Simple Science Series)
I. Title II Gordon, Mike III. Series 552

ISBN 0 7502 1600 X

Typeset by Liz Miller.
Printed and bound in Italy by G. Canale and C.S.p.A, Turin.

Contents

Rocks are made of tiny hard bits joined together in lumps. The tiny hard bits are called minerals. Rocks are not alive.

Soil is a mixture of very small rocks, minerals, bits of dead plants and animals, water and air. Most plants grow in soil.

Soil is all around you. It is under most plants, roads and houses. It is even at the bottom of rivers, seas and lakes. Collect some soil. What colour is it? How does it feel? Are there any rocks or living creatures in it?

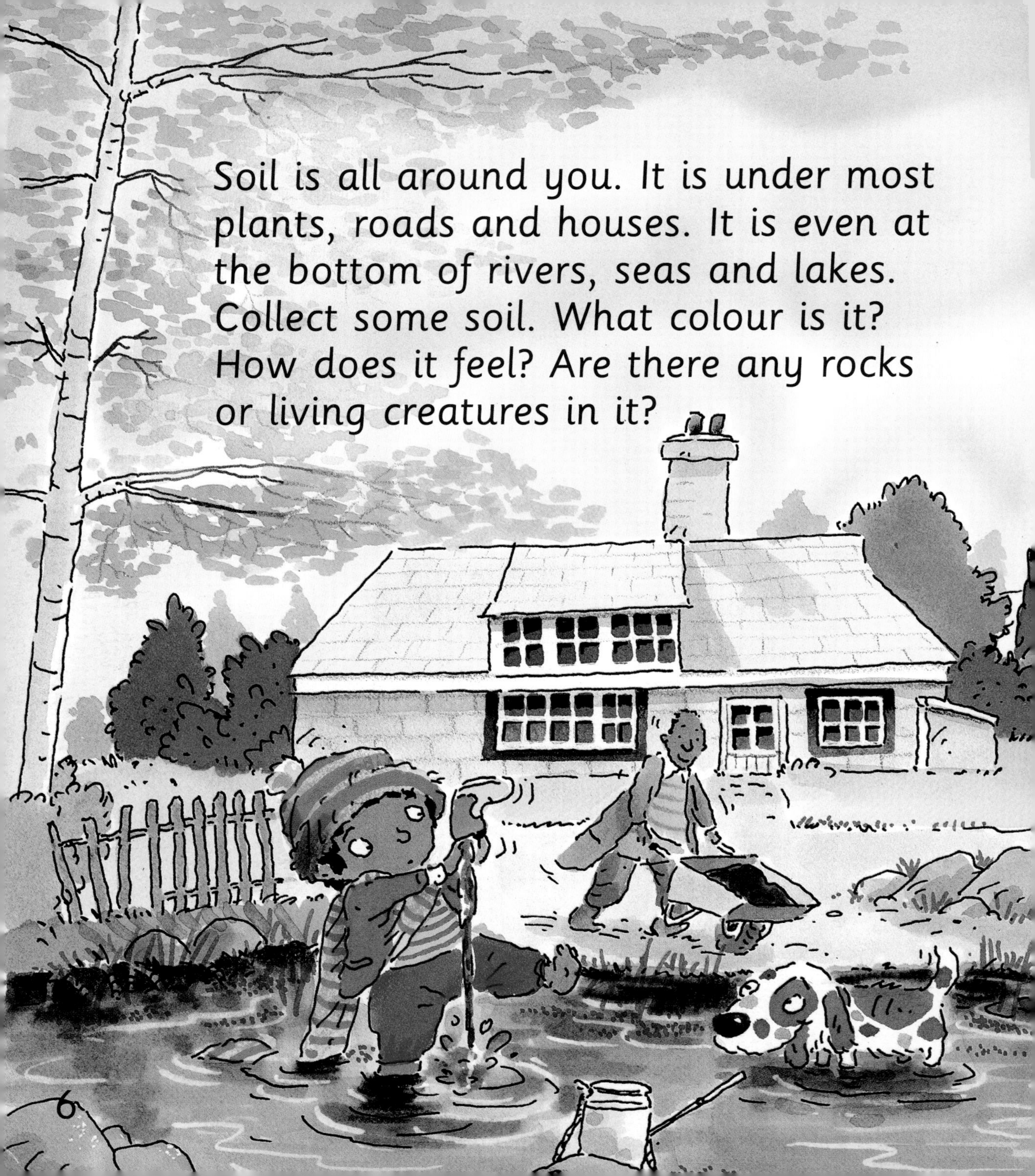

You can see rocks in the ground and in cliffs. People make roads, walls and buildings with them. Find some rocks. Are they big or small? How do they feel? What colour are they?

Long ago, people used rocks to make tools and weapons. People learnt to cook with hot rocks. They made pots out of rocks and built walls and buildings.
Beautiful rocks were worn as jewellery.

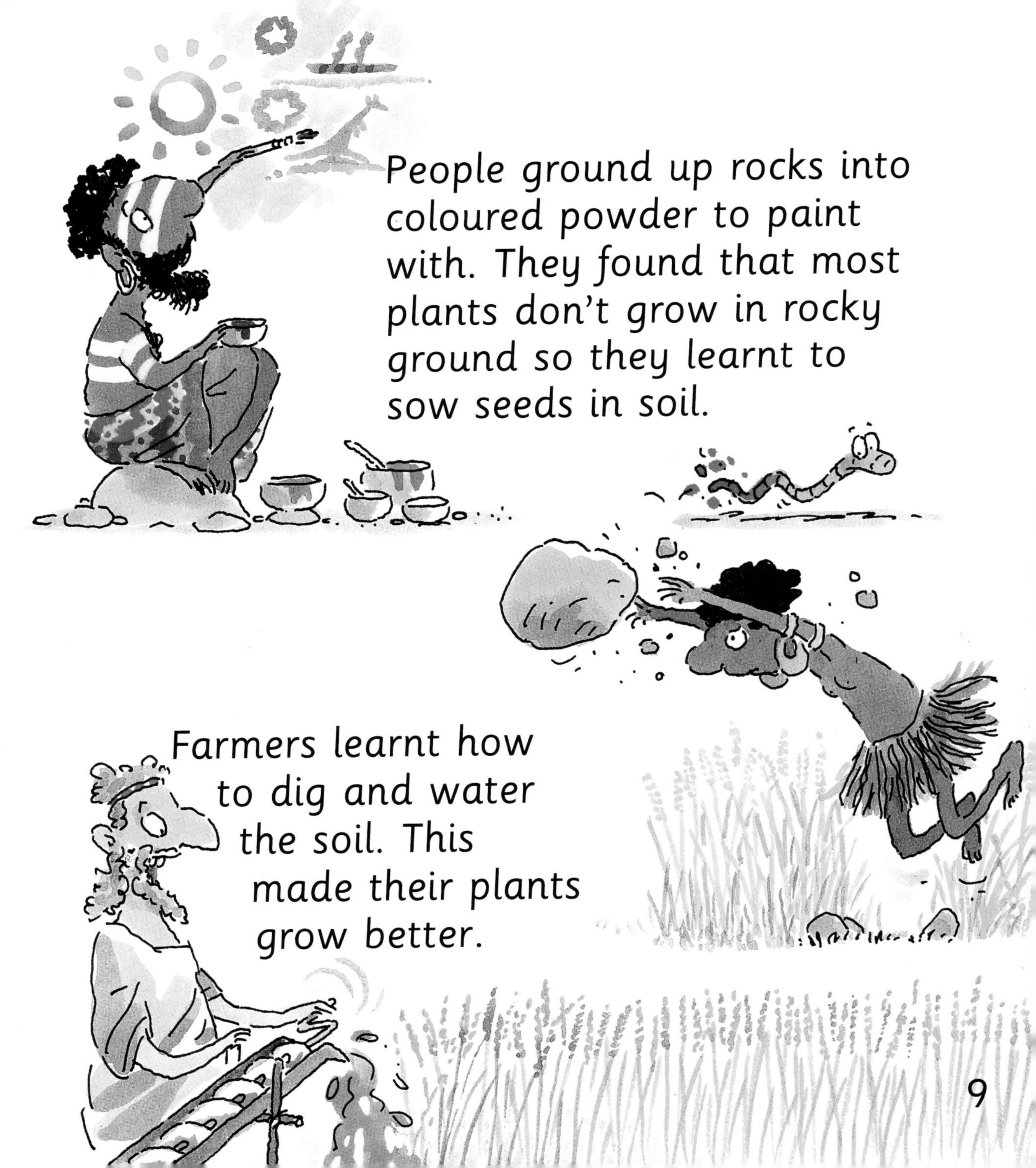

People ground up rocks into coloured powder to paint with. They found that most plants don't grow in rocky ground so they learnt to sow seeds in soil.

Farmers learnt how to dig and water the soil. This made their plants grow better.

Rocks are made
in different ways.
Heat makes
minerals stick
together in lumps.
Volcanoes make
this kind of rock.

Sugar is like a mineral.
Ask a grown-up to help
you melt some - do not
touch! It goes hard when
it cools, just like a rock.

Rivers and seas carry minerals like sand and mud to new places. The minerals get dumped on top of each other.

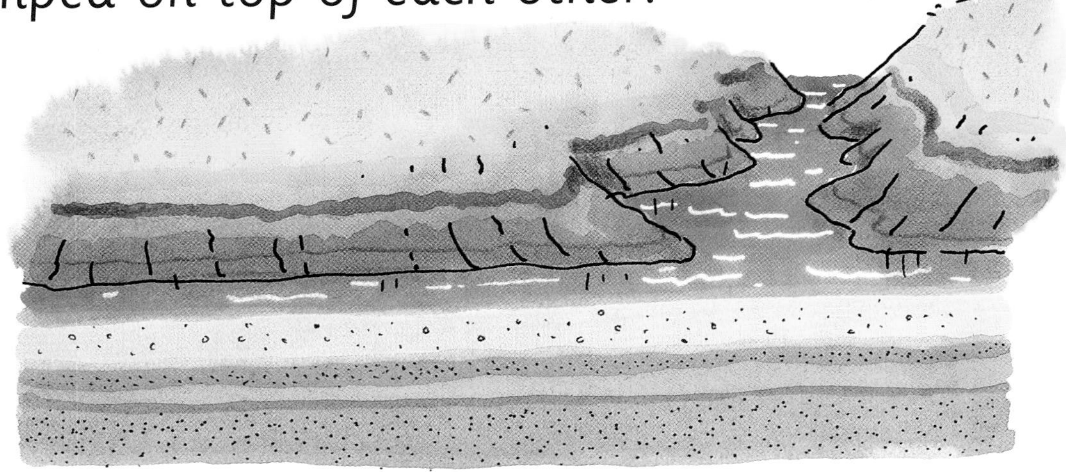

Many minerals slowly turn into rock because they are squashed so hard.

Sugar cubes are made a bit like this. They are grains of sugar squashed together.

Sometimes these minerals fall on animals and plants. The animals and plants are slowly pressed on to the sand and mud. This is how fossils are made. Some fossils are the hard parts of animals and plants that were buried. Others are just the shapes the animals made when they were squashed.

Dead leaves, seeds and branches fall on top of each other. This has happened for thousands of years. Look at some coal. This is made of the plants at the bottom that got squashed and turned hard!

Water can carry minerals. If the water dries up, the minerals get left behind.

Water dripping in caves leaves minerals behind. Very slowly, this makes big lumps of rock.

The flakes inside kettles are minerals.

Wind, sun and rain all make rocks break up into little pieces. This is called weathering.

The sun warms rocks. This makes them a tiny bit bigger.

When the rocks cool off at night, they shrink a little. This heating and cooling makes bits of rock crack off. Look for little stones around big rocks.

Water fills more space when it freezes. Sometimes it freezes inside rocks. The ice pushes against the rocks. This makes them split.

Put some cracked and crumbly rocks inside a plastic tub.

Cover them with water and freeze them. Let the ice melt.

Small pieces of rock may break off into the water. Some cracks will be bigger.

Water rubs away the rough edges of rocks and stones. This makes pebbles and sand. Water carries the pebbles and sand and they help break off big pieces of rock. Waves of water smash away rock, too.

The wind rubs away the softest part of rocks. Plants also break up rocks as they push through any cracks. Even animals can break and crack rocks.

Small pieces of rock help to make soil. Some soils have bigger bits of rock than others. In sandy soil you can see little bits. In silty soils you need a magnifying glass to see them. The bits of rock in clay soils are too small to see.

Next time it rains, see how long it takes the ground to dry. Clay soils go very muddy and take a long time to dry.

Sandy soils dry quickly.

Silty soils dry more slowly than sandy soils, but quicker than clay soils.

Ask a grown-up to help. Find three different looking soils. Cut out the bottoms of three plastic cups.

Put the cut ends of each cup into the three soils. Pour the same amount of water into each cup. The water soaks fastest into the soil with the most sand in it.

Soil builds up on top of rock. Ask a grown-up to help you dig down a few inches in a field. See if you hit rock. Look out for worms and insects. See how the soil looks dark on top and lighter underneath.

The darker part of soil is called topsoil. It has many bits of plants and animals. They are called humus. They are rotting. This means that creatures in the soil are eating them and turning them into smaller and smaller pieces.

Collect three different kinds of soil in tubs.

Put 5 cm of each soil into jars.

Fill the jars with water and wait for ten minutes.

* Keep the tubs of soil - you will need them later!

The biggest pieces of rock sit on the bottom. The tiniest pieces mix with the water and make it cloudy.

Humus floats on top of the water.

Clay soil makes the water cloudiest and silty soil has a lot of humus.

Can you see any little bubbles in the jars? These are filled with air that was in the soil.

Put your tubs of soil in a bright warm spot. Check them each day for a week. Water them if they go dry and look for seeds sprouting and for living creatures.

At the end of the week, plant your own seeds. See which kind of soil they like best.

Plants have grown in the world for millions of years. The first seeds grew in tiny pieces of weathered rock. They fed on the minerals in the rock.

When they died, they made humus. The humus was food for tiny creatures and more seeds. The mixture of rocks and humus made soil. This took thousands of years.

Wind blows soil away. Rain washes it away. This is called erosion. Watch the dust coming from bare soil on a dry, windy day.

Plants help to stop erosion. Pour water down a grassy slope and a bare slope. Grass slows the water down.

The soil on the bare slope mixes with the water and washes away.

Keeping trees, hedges and plants helps to stop the soil eroding.

Soil needs humus to feed the things we grow. People can save grass clippings, leaves and old plants to make humus.

Look at the picture. How can you help to make soil and to keep it safe?

Notes for adults

The 'Simple Science' series helps children to reach Key Stage 1: Attainment Targets 1-4 of the Science National Curriculum.
Below are some suggestions to help complement and extend the learning in this book.

4/5 Tend a rock garden. Draw 'buried treasure' maps. Drag magnets through soil to collect tiny pieces of iron. Collect and test rocks using Mohs scale of hardness tests: crushed by finger nail; scratched by finger nail; scratched by copper coin; scratched by glass etc.

6/7 Visit farms and gardens. Collect and display soil samples.

8/9 Look at sculpture, ancient tools, cave paintings and jewellery. Investigate early irrigation and Archimedes' Screw. Research pyramids, Easter Island and Stonehenge. Make scale models. Use pestles and mortars. Tell the parable of the sower (Luke Ch.8 Vs. 5) and the legend of Jason and the Dragon's Teeth. Research birthstones.

10/11 Use pumice stone (igneous rock). Look at stratification in exposed rock (sedimentary): measure each layer. Grow sugar crystals.

12/13 Visit glacial deposit sites. Go fossil hunting. Make plaster casts. Investigate stones like amber and jet.

14/15 Visit caves. Research gold panning. Plot gold rush sites on maps. Cook meat on rocks heated for three or more hours in a campfire or barbecue. Look at how reptiles use hot rocks.

16/17 Spot weathering on roads, buildings, etc. Compare shapes of rocks and stones with man-made bricks, etc. Display pictures of famous gorges and canyons. Investigate legends, such as Paul Bunyan and the Grand Canyon. Make up new stories about such sites.

18/19 Follow paths worn by animals. Find and measure roots breaking through man-made surfaces. Investigate how the Leaning Tower of Pisa is sinking into clay soil. Display clay and pottery, sand and silicon chips. Research dams, silt and ox-bow bends in rivers.

20/21 Make mud pies and sandcastles - compare how much water is needed. Discover which plants grow best in which soils.

22/23 How much of waste is organic? Start or add to a compost heap. Compare smell of materials rotting in compost heap and airtight container - investigate bacteria. Make a worm farm.

24/25 Investigate soil types of river and pond beds by gently stirring them up. Research rice paddies.

26/27 Compare mountain-side, desert and river valley vegetation. Look at peat and fertilizers. Mulch some soil. Research crop rotation. Visit an archaeological site: show how soil depth helps measure age.

28/29 Display soil erosion pictures from deforested, dammed and monoculture sites around the world. Investigate guano and peat extraction, desertification, soil salination and effects of Amazon gold panning. See how roots bind soil, eg. on sand dunes. Join re-planting schemes.

Other books to read

Geology by G Peacock and J Jessan (Wayland, 1994)

Peterson First Guide to Rocks and Minerals by F H Pough (Houghton Mifflin, 1991)

Rock Collecting by Roma Gans (A & C Black, 1990)

Science for Kids: 39 Easy Geology Experiments by R W Wood (Tab Books, 1991)

Index